CHEMISTRY MATH BASICS

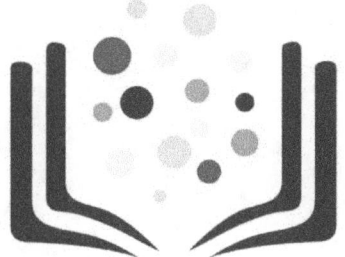

Mody Educational Resources

EFFECTIVE BITE-SIZED LEARNING FOR THE MODERN STUDENT

High School Chemistry: Chemistry Math Basics

<u>Other Study Aids by Mody Educational Resources</u>

SAT Series
Rock the SAT Math Test: 6 Full Length Math Tests
Rock the SAT Math Test: 4 Difficult Math Tests

High School Biology Series
High School Biology: The Cell
High School Biology: The Chemistry of Life
High School Biology: Cell Division
High School Biology: Genes and Heredity
High School Biology: Digestion and Nutrition
High School Biology: The Circulatory System
High School Biology: The Respiratory System
High School Biology: Plants: Form and Functions
High School Biology: Taxonomic Classification Systems
High School Biology: The Invertebrates

High School Physics Series
High School Physics: Newton's Laws of Motion
High School Physics: Projectile Motion
High School Physics: Translational Motion
High School Physics: Conservation of Momentum
High School Physics: Energy and Equilibrium

High School Chemistry Series
High School Chemistry: Chemistry Math Basics
High School Chemistry: Atomic and Molecular
Structure
High School Chemistry: Chemical Bonds
High School Chemistry: Stoichiometry
High School Chemistry: Acids and Bases

High School Chemistry: Gases
High School Chemistry: Solutions
High School Chemistry: Rates of Chemical Reactions
High School Chemistry: Chemical Equilibrium
High School Chemistry: Thermodynamics and Physical Chemistry
High School Chemistry: Introduction to Organic Chemistry
High School Chemistry: Introduction to Nuclear Chemistry

Visit the following sites for additional resources:

www.modyeducation.com where you will find tons of great advice, tips, and resources.

www.exammasters.ca/blog where Dr. Vishal Mody offers awesome tips, advice, and free study resources.

Acknowledgements

This Chemistry booklet is the culmination of years of tutoring Grade 11 and 12 high school chemistry students. With over 10 years of tutoring experience, I have clearly understood where students have trouble with particular topics in Chemistry and have endeavoured to create a resource that will make specific topics easier to learn. Many thanks to all my students, past and present, for challenging me to be the best I can be.

Enjoy the process of learning,
Dr. Vishal Mody

Authors Note

Many Grade 11 and 12 students have a lot of trouble with concepts in Chemistry. These students often are not taught specific concepts well and without that solid foundation, it becomes very hard to understand the more difficult concepts.

This chemistry booklet was created to help students specifically with the basics of math that they will have to know in order to understand and perform well in this subject. This booklet has been made extremely concise yet explains the concepts in detail at the same time. Also, this booklet is not designed to be your main study source, but rather, as an adjunct to your school teacher's notes. There are also lots of practice questions with detailed solutions at the end to solidify the concepts you have learned.

Best wishes in your studies,
Mody Educational Resources

Contents

Accuracy and precision

Accuracy is defined as the closeness of measured values to the actual or accepted value.
Precision is defined as the closeness among measured values of some quantity.

Example

The standard value of length of an object is 50 cm (actual value). Two students recorded its length five times.

1. The first student recorded the length as 52, 47, 50, 51 and 49 cm.
2. The second student recorded measurements as 44, 45, 45, 44 and 44 cm.

In this case, the measurements recorded by first student are more accurate (because they are closer to the actual measured value) but less precise. The measurements of the second student are more precise (because they are closer to each other) but less accurate since they are not close to the actual value.

The accuracy of a measurement is the direct measure of fractional or percentage uncertainty in that measurement. A more accurate reading has less fractional or percentage uncertainty.

The precision of a measurement depends upon the instrument that is being used. Therefore, precision is a direct measure of least count (absolute uncertainty) of that instrument. A reading is said to be more precise if the least count or absolute uncertainty of instrument is less.

$$\text{Fractional uncertainty} = \frac{\text{absolute uncertainty}}{\text{measured value}}$$

$$\text{Percentage uncertainty} = \frac{\text{absolute uncertainty}}{\text{measured value}} \times 100\%$$

QUESTIONS

1. A length of a straight wire is measured as 32.5 cm with the help of a meter rod having least count of 0.1 cm. Another measurement of length of an object is recorded as 0.63 cm by using Vernier Caliper having least count 0.01 cm. Which one of the measurements is more precise and which one is more accurate?

2. Calculate the percent uncertainty in a measurement of length taken by micrometer screw gauge as 0.403 cm.

Significant figures

Significant figures comprise of all the accurately known digits along with the 1st doubtful digit in any measurement.

In the measurement of any physical quantity there is an inevitable uncertainty associated with it. One reasonable cause of this uncertainty is the instrument that we use to measure any physical quantity. Every instrument of measurement is calibrated up to a definite smallest division. For example, the meter rod is calibrated in millimeters (mm) and has a least count of 1 mm (or 0.1 cm).

Example

We measure the length of a straight rod (with a ruler) and we record the measurement as 16.4 cm. Here, the digit 1 and 6 are accurately known while digit 4 is the estimated one because its value may lie between 16.35 cm and 16.45 cm since the smallest division on the ruler is 0.1 cm. Therefore, there are 3 significant figures in this measurement. However, if the same measurement is taken with the help of a Vernier caliper (picture below), which can measure accurately up to a hundredth of a centimeter, the recorded value would be 16.40 cm having 4 significant figures and by using the micrometer screw gauge our recorded value would contain up to 5 significant figures.

So, by using better instruments with smaller divisions of measurement, we can improve the accuracy of our measurement. This also means that our measurements would have more significant figures.

Here's what a Vernier caliper looks like:

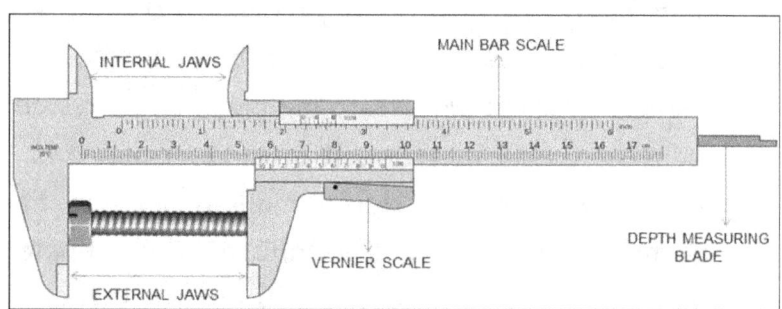

Rules for counting significant figures

1. All of the non-zero digits 1, 2, 3, 4, 5, 6, 7, 8, 9 are significant.

   Example: 2254691 = 7 significant figures

2. Zeros may or may not be significant. In the case of zeros:

   - All the zeros present between 2 significant figures are counted.

     Example: 90078 = 5 significant figures; 3.09 = 3 significant figures

   - The zeros that lead the significant figures are non-significant.

     Example: 0.00067 = 2 significant figures; 06.92 = 3 significant figures

   - The zeros that trail significant figures in a decimal fraction are significant.

     Example: 5.800 = 4 significant figures

- The number of significant trailing zeros in integers depends upon how accurately the instrument measures.

  Example: The weight of an object is 40000 kg, which is 5 significant figures if the smallest division on the measuring scale is 1 kg. It will be written in scientific notation as 4.0000×10^5. However, if the smallest division of the measuring scale is 10 kg there will be 4 significant figures and it will be written as 4.000×10^4.

3. In scientific notation all the figures are significant else than the power of 10.

   Example: 9.08×10^4 = 3 significant figures

Rules for calculating with significant figures

1. Addition or subtraction

Here, the position of decimals matter rather than the number of significant figures. When adding and subtracting numbers, you have to look at the amount of significant numbers in the decimal portion of each value. You then round your answer to the least amount of decimal numbers.

Think about it this way:

Imagine you and 3 of your friends have to finish a race **together**. How fast you all finish the race will depend on the slowest person, not the fastest. Similarly, when adding/subtracting values, how precise you are is determined by the least precise value (which has the least amount of decimal numbers). You can't be more precise than the least precise value!

Examples:

- 82.3 + 2.16 + 0.007 = 84.467 → rounded as 84.5
  Here, 82.3 has the least number of decimal places.

- 3.6132 + 1.21 + 4.316 = 9.1392 → rounded as 9.14
  Here, 1.21 has the least number of decimal places.

- 42.00 + 13.050 → rounded as 42.05
  Here, 42.00 has the least number of decimal places.

2. Multiplication or division

In the case of multiplication and division, the product or quotient is rounded to match the value with the least amount of significant figures (the least accurate number). This is similar to when we add or subtract, however, in this case we are taking all the significant figures in the **entire** number into consideration, as opposed to just in the decimal part of the number.

Examples:

- $3.\underline{5} \times 9.54 = 33.39 \rightarrow$ rounded as 33.4
  The number 8.57 has the least number of significant figures.

- $7.314 \div (6.851 \times 10^{-3} \times \underline{8.57} \times 10^3) =$
  $8.027491112934099 \rightarrow$ rounded as 8.03
  The number 8.57 has the least number of significant figures.

QUESTIONS

3. How many significant figures are there in:
    A. 0.001010
    B. 04.120
    C. 6.009

4. Calculate the number of significant figures in numbers expressed in scientific notation:
    A. 9.100×10^4
    B. 7.03×10^5

5. Suppose that the least count of a weighing machine is 10 kg. Find out the number of significant figures in the following values and also express them in scientific notation.
    A. 692000 kg
    B. 17000000

6. Give the answer in significant figures when we added following four quantities: 9.8, 0.0004, 1.89 and 3.008.

7. Find the volume of an object having length of 7.834, width of 0.0723 and a height of 3.23 in cm. Give the answer in scientific figures.

Scientific notation

Scientific notation is the standard form in which numbers are expressed in powers of ten.

Example: 259.8 is written as 2.598×10^2

Usually, this is done for very large numbers that are difficult to work with in the standard way we write numbers. So, for the sake of convenience the numbers are written in a short form by employing powers of ten.

> **Think about it this way:**
>
> If you had to add 1 trillion to 1 googol, would you want to write out all the zeros? Probably not! It would be way easier to simply write it as 1.0×10^{12} + 1.0×10^{100}.

Rules for conversion

1. From actual numbers to scientific notations

   A. It is an international practice to write only 1 non-zero digit to the left before placing the decimal point.

B. If the decimal point is shifted towards the left the exponent is written as positive.

Example: 127.39 is written as 1.2739×10^2

C. If the decimal point is shifted towards right the exponent is written as negative.

Example: 0.0032 is written as 3.2×10^{-3}

2. From scientific notations to actual numbers

> **This is opposite of rule 1! Just memorize rule 1 and do the opposite when you have to go from scientific notation to the actual number.**

A. In the case of a negative exponent, the decimal is shifted towards the left.

Example: 1.02×10^{-2} will be written as 0.0102

B. If the exponent is positive, the decimal is shifted towards right.

Example: 6.3×10^4 will be written as 63000

QUESTIONS

8. An object weighs 954006.8 kg. Express this weight in scientific notation.

9. What will the length of a rod in millimeters be, expressed in scientific notation, if its length is recorded as 6.54 meters?

Physical constants

Physical constants refer to the physical quantities that are considered as universally constant in nature and time. Meaning that they always remain the same!

Certain fundamental quantities appear again and again in nature and so, they are used in the basic theoretical equations of physics. These special numbers are denoted by universal and specific symbols. In order to check the correctness of the theories these physical constants must be known to a very high accuracy.

Commonly used physical constants

Symbol	Name	Value	Unit
c	Speed of light in a vacuum	299,792,458	m/s
g	Gravitational constant	6.67384×10^{-11}	N•m^2/kg^2
h	Planck's constant	$6.6260695 \times 10^{-34}$	J•s
e	Elementary charge	$1.602176565 \times 10^{-19}$	C
m$_e$	Electron mass	$9.10938291 \times 10^{-31}$	kg
m$_p$	Proton mass	$1.672621777 \times 10^{-27}$	kg
m$_n$	Neutron mass	$1.674927351 \times 10^{-27}$	Kg
N$_A$	Avogadro's constant	$6.02214129 \times 10^{23}$	1/mol
K	Boltzmann's constant	$1.3806488 \times 10^{-23}$	J/K
R	General gas	8.3144621	J/mol•K

constant

QUESTIONS

10. For any two physical constants, define them and state their name, symbol and value.

11. State why the physical constants are measured with great accuracy?

International system of units

In 1960, the international system of units (SI) was established. The SI system set the definitions and standards for the particular units of the physical quantities. At present it is the most popular and accepted system of units. In SI system, 2 major classes of units are described.

Base units

Base units are assumed to be mutually independent. For various physical quantities the SI system has presented 7 base units:

Physical Quantity	SI Unit	Symbol	Dimension Symbol
Length	meter	m	L
Mass	kilogram	kg	M
Time	second	s	T
Electric current	ampere	A	I
Thermodynamic temperature	kelvin	K	Θ
Amount of substance	mole	mol	N
Luminous intensity	candela	cd	J

Definitions of base units

Length (meter)

The meter is the length of the path travelled by light in vacuum during a time interval of 1/299, 792, 458 of a second.

Mass (kilogram)

The kilogram is the unit of mass; it is equal to the mass of the international prototype of the kilogram.

Time (second)

This may actually come as a shock to you or you have never really thought about it. But, the second is derived from the duration of 9 192 631 770 periods of the radiation corresponding to the transition between the two hyperfine levels of the ground state of the cesium 133 atom. This is way too much info and you don't have to know it! But it's a great fact to impress someone with.

Electric current (ampere)

The ampere is that constant current which, if maintained in two straight parallel conductors of infinite length, of negligible circular cross-section, and placed 1 meter apart in vacuum, would produce between these conductors a force equal to 2×10^{-7} newton per meter of length.

Thermodynamic temperature (kelvin)

The kelvin, unit of thermodynamic temperature, is the fraction 1/273.16 of the thermodynamic temperature of the triple point of water.

Amount of substance (mole)

The mole is the amount of substance of a system which contains as many elementary entities as there are atoms in 0.012 kilogram of carbon 12; its symbol is "mol." When the mole is used, the elementary entities must be specified and may be atoms, molecules, ions, electrons, other particles, or specified groups of such particles.

Luminous Intensity (candela)

The candela is the luminous intensity, in a given direction, of a source that emits monochromatic radiation of frequency 540×10^{12} hertz and that has a radiant intensity in that direction of 1/683 watt per steradian.

Derived units

Derived units comprise of all the units other than the base units that can be algebraically expressed on the basis of base units or other derived units. The symbols of derived units are obtained by either multiplying or dividing the base units.

Example: Area and volume are derived quantities having units of meter squared and meter cubed, respectively.

Derived Quantity	Unit	Symbol	In terms of base units
Frequency	Hertz	Hz	s^{-1}
Force	Newton	N	$kg{\bullet}ms^{-2}$
Work	Joule	J	$kg{\bullet}m^2s^{-2}$
Power	Watt	W	$kg{\bullet}m^2s^{-3}$
Pressure	Pascal	Pa	$kg{\bullet}m^{-1}s^{-2}$
Electric charge	Coulomb	C	$A{\bullet}s$

Metric prefixes

Metric prefix refers to a unit prefix that is placed before the basic unit to indicate a fraction or multiple of that base unit. Metric prefixes are applied when there is a need of converting very small or large numbers into numbers that are more convenient to work with.

A List of the Metric Prefixes

Prefix	Symbol	Exponential
Yotta	Y	10^{24}
Zetta	Z	10^{21}
Exa	E	10^{18}
Peta	P	10^{15}
Tera	T	10^{12}
Giga	G	10^{9}
Mega	M	10^{6}
Kilo	k	10^{3}
Hecto	h	10^{2}
Deca	da	10^{1}
Deci	d	10^{-1}
Centi	c	10^{-2}
Milli	m	10^{-3}
Micro	μ	10^{-6}
Nano	n	10^{-9}
Pico	p	10^{-12}
Femto	f	10^{-15}
Atto	a	10^{-18}
Zepto	z	10^{-21}
Yocto	y	10^{-24}

Rules for indicating units and prefixes

1. The complete name of a unit will never start with a capital letter even if it is name of scientist (i.e., kelvin, newton, etc.).
2. The unit that is named after a scientist is expressed with a symbol having the first letter capitalized (i.e., N for newton and Pa for pascal).
3. When 2 base units are written with each other, space is used between them such as N m or as N•m to show that they are a product.
4. Two or more of the same prefix cannot be used simultaneously within a single unit. E.g., 1nnF (nano) is written as 1aF (atto).

QUESTION

12. Define SI base units and also state their names, symbols, units.

13. State the standard definition of second and mole.

14. How can force, work and power be defined in terms of base units?

15. How many picometers are there in 100 micrometers?

16. The length of a rod is 4 meters. Convert it to micrometers.

Temperature conversions

TEMPERATURE SCALES

A temperature scale is a scale that gives us the quantitative measurement of temperature. At present, three temperature scales are more generally used for different purposes: Celsius, Kelvin and Fahrenheit.

CELSIUS

Celsius temperature scale is also known as centigrade temperature scale. It was invented by the Swedish astronomer Anders Celsius in the year 1742. The Celsius scale has 100 divisions between two defined points. It is based on zero degrees for the freezing point of water and a hundred degrees for the boiling point of water. The Celsius scale is commonly used.

KELVIN

In the International System (SI) of measurement, the kelvin temperature scale is the base unit of thermodynamic temperature. It is defined as 1/273.16 of the triple point of pure water. The triple point refers to the point where the solid, liquid, and gaseous phases of water are at equilibrium. This scale has 100 divisions just like the Celsius scale. However, the Kelvin scale defines absolute zero (the theoretical temperature at which molecules possess the lowest energy) as zero degrees. In the Kelvin scale, the freezing point of water is 273.16 degrees and the boiling point is 373.16 degrees. By using the Kelvin scale, several physical laws and derivations can be expressed more simply.

FAHRENHEIT

Fahrenheit temperature scale is divided into 180 parts. It is based on 32 for the freezing point of water and 212 for the boiling point of water.

CONVERSIONS

From	To Fahrenheit	To Celsius	To Kelvin
Celsius (C)	(C * 9/5) + 32	C	C + 273.15
Kelvin (K)	(K - 273.15)* 9/5 + 32	K - 273.15	K
Fahrenheit (F)	F	(F - 32) * 5/9	(F - 32) * 5/9 + 273.15

QUESTIONS

17. Convert the absolute zero (0 K) into Fahrenheit.

18. A patient is suffering from a high fever of 104° F. Convert it into Celsius and state how much his body temperature is increased if the normal body temperature is 37° C.

19. Convert the boiling point of water from the Celsius scale to kelvin and Fahrenheit.

Experimental errors

No physical quantity can be measured to perfection. There is **always** some uncertainty or impreciseness, which is associated with physical measurement. This is known as experimental error and is defined as the difference between the theoretical and experimental value. The theoretical value refers to the actual or known value while the experimental value is the calculated or measured value.

Types of experimental errors

1. Random error

A random error affects the precision of the measurement. In the case of a random error, different values are obtained in spite of repeating the measurements under the same conditions. They occur due to unknown reasons. These errors are rectified by taking measurements many times and then calculating the average value of all the measurements.

2. Systemic error

A systemic error affects the accuracy of the measurement. In a systemic error, the measured values differ by the same amount from actual values under repeated measurements. A consistent difference is observed in the results. It can occur due to incorrect marking, faulty calibration or zero error. This type of error can be rectified by comparing the faulty instruments with the standard or accurate instrument and then applying the suitable correction factor.

Calculating experimental error

The formula used for calculating percentage error in an experiment is given as:

$$\% \ Error = \frac{E - A}{A} \times 100$$

Where: E = experimental value; A = actual value

QUESTIONS

20. Differentiate between the two types of experimental errors.

21. A student experimentally determines that the diameter of a sphere is 2.76cm. The actual value is 2.92cm. Calculate the percent error in this measurement.

Cancelling units

In solving physical problems, there are several occasions when we have to convert between different measurement units. For example, let's say that we have measured the speed of a car in feet per minute by measuring the distance in feet and the time in minutes. But what if we want it to be in standard form (i.e., miles per hour)?

Conversions are among the most important tasks performed while solving problems. So it is wise to approach a conversion in a systematic way. We can easily cancel out the units that we don't want by multiplying the quantity with whose value is one.

Example

Convert 5 meters into centimeters.

STEP 1: start with the given quantity = 5m

STEP 2: quantity to be measured = cm

STEP 3: create conversion factor equal to one

$$= \frac{1m}{100cm} \text{ or } = \frac{100cm}{1m}$$

STEP 4: select and multiply the conversion factor to cancel the unit

$$= 5\cancel{m} \times \frac{100cm}{1\cancel{m}} = 500 \text{ cm}$$

Notice that the meter units cancel out!

QUESTIONS

22. How many inches are in 9.8 meters?

23. How many seconds are in 84 minutes?

ANSWERS

1. The reading measured with a Vernier caliper that has a least count of 0.1 cm is more precise than that of 0.01.
2. 0.00248%
3. (A) 4 (B) 4 (C) 4
4. (A) 2 (B) 3
5. (A) 5 (B) 7
6. 14.7
7. 1.83
8. 9.54×10^8
9. 6.5×10^3
10. Physical constants refer to the physical quantities that are considered universally constant in nature and time.

    h - Planck's constant $6.6260695 \times 10^{-34}$ J•s

    e - Elementary charge $1.602176565 \times 10^{-19}$ C

11. In order to check the correctness of the theories, these physical constants must be known to a very high accuracy.
12. At present it is the most popular and accepted system of units. In SI system, 2 major classes of units are described.

13. Time (second): The second is the duration of 9 192 631 770 periods of the radiation corresponding to the transition between the two hyperfine levels of the ground state of the cesium 133 atom.

Amount of substance (mole): The mole is the amount of substance of a system which contains as many elementary entities as there are atoms in 0.012 kilogram of carbon 12; its symbol is "mol." When the mole is used, the elementary entities must be specified and may be atoms, molecules, ions, electrons, other particles, or specified groups of such particles.

14. Force = Newton = $kg \bullet ms^{-2}$

It is defined as a force that would give a mass of one kilogram an acceleration of one meter per second per second.

Work= Joule = $kg \bullet m^2 s^{-2}$

Work is equal to the work done by a force of one newton when its point of application moves one meter in the direction of action of the force.

Power = Watt = $kg \bullet m^2 s^{-3}$

It is equivalent to one joule per second.

15. 10^8 Pico meters

16. 4.0×10^6
17. -459.67
18. 40 and temperature rises by 3 C.
19. 373.15 K & 212 F
20. Random error affects the precision of the measurement while systemic error affects the accuracy of the measurement.
21. 5.479%
22. 385.827 inch
23. 5040s

www.ingramcontent.com/pod-product-compliance
Lightning Source LLC
Chambersburg PA
CBHW070339190526
45169CB00005B/1961